I0390815

Catchments and Carbon

Catchments and Carbon

The real cause of
unstable weather

Lois Caborn

CABORN BOOKS
ENGLAND

First Edition January 2023
Copyright © Lois Caborn 2023

The moral right of the author has been asserted.

All rights reserved.
*No part of this publication may be reproduced, stored in a
retrieval system, or transmitted, in any form or by any means,
without the prior permission in writing of the publisher, nor be
otherwise circulated in any form of binding or cover other than that in
which it is published and without a similar condition including this
condition being imposed upon the subsequent purchaser.*

*Typeset in Palatino-Roman,
and Arial*

ISBN: 978-1-4478-0424-6

Caborn Books, England

Contents

Carbon is the basis of life.
Water allows the process of life to continue.
Remove carbon and you remove water.
Remove water and you remove life.

Say NO to Net Zero!

Preface

The purpose of this book is to bring together many years of research and implementation in the field of what can loosely be described as earth sciences.

The information provided is done so in order to demonstrate a coherent structure to the hypothesis that the "global warming scenario" presently being proposed is in fact a fake and that we have instead a situation of climate instability.

The industrialized methodology of farming, together with large scale deforestation, have accelerated a process that has been growing in magnitude for generations. This process is the removal of water from the land and the drying out of vast acreages on a global scale.

Drying of the land allows the oxidation of the soil carbon and this is in reality the major source of carbon dioxide that allegedly produces global warming. In fact the removal of carbon from soil renders it incapable of holding and storing water.

It is this lack of soil water that is producing the temperature increases being experienced not the slight increase in carbon dioxide in the atmosphere.

Water provides the heat transfer mechanism for the planet. The evaporation of water from organic soils brings about a cooling of the ecosystem. The water vapour rises, forms clouds, and returns downwind as rain.

Present agricultural soils are becoming dry and dessicated and no longer act as reservoirs for rainwater.

A further major difficulty is the huge deforestation occurring in many parts of the world. Sixty percent (60%) of terrestrial rainfall is produced by evaporation of water by trees and the soil they grow in. As soils dry and deforestation spreads, rainfall is depleted, the cooling effect is lost, and temperatures increase.

It is imperative that carbon and water are returned to the land and that catchments are managed on a human scale by the people living in them.

The sequestration of carbon and the other key nutrients nitrogen and phosphorous can be achieved by working with nature and in a way that is not possible on the scale of "big infrastructure" that seeks to maximize corporate profit rather than human wellbeing.

Predicted shortages in water for drinking and irrigation can be avoided by reversing the loss of water from the land and managing catchments properly.

The call for construction of new reservoirs is simply a way of creating "asset banks" of essential water supplies isolated from the people and their ecosystems although the people will have provided the "subsidy" through taxation for their construction and will pay through metering for consumption. The price will of course keep on increasing as the planet heats up.

The statement: "due to global warming" is now used as the catch all for every major and minor catastrophe from environmental refugees to forest fires without any real explanation of how this comes about.

The solution always appears to be the annihilation of coal, oil and gas industries in a way that would prevent their return when the new paradigm was eventually proved to be false.

Hostile actions from Russia have shown how delusional this policy is and the vulnerability of the UK, which has more of these resources than most, is a case study in criminal mismanagement.

In explaining the linkages between soil, water and temperature and such things as the impact of housing on catchments, the author seeks to offer a more rational explanation to the climate instability being experienced.

In doing so, the problem of carbon sequestration into soil and the methodology of detoxifying aquifers from phosphorous and nitrogen are also covered and above all the primary requirement of returning water into and onto the catchments is covered by detailed examples proven to be effective and available to all in a partnership with nature.

Chapter 1

General Considerations

Sometime after the last ice age, 20,000 years ago, a new activity appeared on Earth. Humanity, from being hunter gatherers, transitioned to farming.

As a consequence forests were burned for clearings in which crops were planted. Burning wood of course adds carbon dioxide to the atmosphere and additionally, the movement of carbon from the soil into cereals to be consumed as food added a new part to the equation that expresses the mass balance of the carbon cycle.

Up to this point in time the carbon cycle had achieved a general equilibrium. Carbon dioxide and nitrogen from the air were sequestered into the soil. Soil organisms converted these basic building blocks into nutrients for trees and other plants.

The plants, driven by photosynthesis, pumped the essential water from rainfall out of the soil and back into the sky where it returned to ground again as rainfall. The critical part however was the presence and movement of water in the cycle. Without water all biological processes cease.

By removing carbon from the soil its ability to hold water was, inadvertently, lost. With that loss the green agricultural areas began to recede and desert began to appear.

It is critical therefore to consider how "soil" reacts to the various activities undertaken by man.

The microbial balance in soil respiration was built up over millions of years and essentially consists of cycling carbon, as carbon dioxide, and oxygen into and out of the soil, driven by the presence of a multiplicity of soil organisms including bacteria and fungi who live and breathe within the soil.

Plants of course, as well as employing photosynthesis to manufacture their food for growth, also pump water through their roots and along their stems into the leaves and out into the atmosphere to return as rain.

At the end of their life cycle when plants die, their roots, stems and leaves are transported back into the ground largely by worms where the many types of fungi specializing in decomposition of organic materials convert them into nutrient and humic fibres.

The presence of this decayed material visible as humic or organic matter in the soil, is the organic content that enables the moisture level in the soil to be maintained.

Without this humic material the sand or sandy soil remaining has very little water retention whilst in comparison the organic "peaty" moorland soils retain large amounts.

Within the organic soil there exists a world of soil organisms all of which play their part in retaining a healthy soil. Examination of a soil shows that at the surface there appear to be granules or agglomerates of soil. These agglomerates are formed by a polysaccharide glue manufactured by bacteria in the soil. Their purpose is to protect the (soil) surface from breaking up under the impact of rainfall and to maintain an open porous structure allowing for gas diffusion.

These aggregates are susceptible to break up under the impact of high velocity rainfall.

As weather becomes more unstable there is a tendency to shorter duration, but more intense storms. In this circumstance the raindrops have a higher velocity and greater diameter.

The energy contained in a large raindrop from an intense rain event may be several hundred times greater than that from the drop size and velocity of gentle rain.

The impact from a high energy raindrop will shatter a soil aggregate and as a consequence the tiny fibres of decayed organic matter which give structural stability will be liberated to float off in the surface water.

The overall effect of the breakdown of aggregate and the structural stabilization produced by fibre is to reduce the soil to fine particles. On a field that is sloping these small light particles wash down gradient as a slurry into the nearest drainage channel to the river thereby reducing the flow carrying capacity of the river.

As the water recedes and the field dries, the fine particles of soil remaining will settle and be pulled into the top layer of the soil by gravity and suction of the falling water table. This effectively reduces permeability and access to the aquifer beneath. We then see a "panned" soil that is impermeable and devoid of organic life.

To counter this situation nature has its own engineering division, the worms.

The notable scientist, Charles Darwin, estimated that the earthworms in the pastures adjacent to his house moved 50 tons per hectare per year of soil to the surface of the pasture. This amounts to a layer of top soil approximately half a centimeter thick and equates to creating a new soil layer 15cm thick within 30 years.

This benevolent activity by earth worms could provide an answer to restructuring our damaged pastures, were it not for man's unintentionally malevolent input.

Grazing dairy cattle are susceptible to enteric worms in their gut. These worms grow by taking nutrient from the cow, a situation that obviously needs correcting. This is where anti-worm (worming) agents come in.

Modern worming agents are very effective. One can always tell that you are in a field where the cattle have been wormed. The digestion is affected and once substantially solid cow pats are now semi-liquid. The worming agent can move out of this liquid down into the top soil or can be leached out by rainfall and carried to the nearest water courses. The result is very simple to express, the worms and associated fauna are also terminated. Furthermore, the worming agents are persistent and having moved from the field into the water courses they immediately eradicate the aquatic fauna.

This inadvertent destruction of soil organisms and aquatic life is more than well matched by an even more deliberate action that produced a somewhat spectacular result. As indicated previously the agglomeration and stabilization of topsoil by bacteria can be overcome by simpler methods.

Spraying pastures with an all-round treatment weed killer and pesticide kills virtually all forms of life in the topsoil, including the bacteria that provide the glue.

A recent news bulletin showed the unintended consequences of high energy rainfall impacting on agricultural soil treated in this manner. The mud slurry that resulted had a devastating effect on the small village down gradient of the agricultural area.

These mechanisms of aggregation, structural organic fibre and (worm) engineering are all part of the primary soil structure and they provide a permeability to the top layer of soil allowing air to penetrate and the soil to breathe. Movement and storage of water in this layer is also facilitated both upwards by evaporation into the air (and hence to provide rainfall elsewhere) and downwards into the aquifer.

A grassland or planted soil has a further benefit in that a root system, quite shallow for terrestrial, much deeper for wetland plants, imparts a secondary stabilized structure to the soil and prevents soil erosion and wash off.

The modern day agricultural practice of preparing a field by spraying with weed killer and then light furrow planting of a crop allows this secondary stable soil structure to develop in the growing season, but is removed at harvest time. The soil in this case is acting only as a support for the crop and takes very little part in its nutrient cycle, all nutrient being provided by fertilizer, very little coming naturally from the soil.

The mechanical practice of this method of farming also opens up the topsoil and allows oxidation of the soil carbon. At present some five tons per hectare per annum of carbon is being lost equivalent to approximately eighteen tons of carbon dioxide.

It is already known but not well publicized that this methodology renders modern foodstuffs deficient in many of the mineral elements that food from organic soil grown crops have.

It is also evident that within a few days of sunshine in the UK the field crops are suffering through lack of water. It is also very easy for a helicopter to spot an illegal hosepipe used in a garden during a drought as the green grass is very visible.

This scenario would not happen if the soil organic matter was still present in the upper layers of the field. Retention of several hundred tons per hectare of rainfall would be quite normal and would provide many days watering for any crop.

It is regrettable that water vapour, instead of being retained in the soil, rapidly joins the emissions of carbon dioxide and possibly methane in the list of greenhouse gases.

In discussing climate change or, more correctly, climate instability, much is made of the greenhouse gases carbon dioxide and methane.

Water vapour is seldom if ever discussed and yet it is by far the biggest moderator of climate than the former two. In a hot dry desert the most notable absence is that of clouds. Whilst the overwhelming presence is that of sunshine.

A typical day in England is often noticeable for clouds, a lack of sunshine and low temperatures. Under normal observation therefore it is water that acts as the moderator of temperature and climate.

That this is so should come as no surprise since water

is the key component in the earth's heat transfer system. Water has the largest heat capacity of all known liquids and its presence as solid (ice), liquid (water) and vapour (clouds) will determine the climatic state of a region.

A research chemist is able use solid carbon dioxide as a window through which to pass infra-red radiation since the degree of adsorption experienced is negligible compared to traces of moisture which would obliterate the beam.

If we analyse the atmospheric concentrations of these gases, we see that water vapour by far outweighs the carbon dioxide:

Carbon dioxide 0.04%, methane 1009 parts per billion, water up to 4%.

Nevertheless the policy put forward to cope with climate change is based entirely on the reduction of atmospheric emissions specified as carbon dioxide. It is claimed that a 50% reduction in carbon dioxide emissions is required to prevent a catastrophic 2 degrees rise in the temperature of the planet's climate.

We could perhaps at this point remind ourselves of the early 1980s when we were told of the disaster scenario of global cooling, when rivers and areas of the sea would freeze. Gas and oil were on the point of running out and the only way to supply energy to keep us warm was by going nuclear.

This policy may have worked since large amounts of subsidy (taxation) on nuclear power stations did appear to prevent the seas freezing over. Today we face the odd situation that we have the opposite crisis that oil and gas are plentiful and the climate is running hot.

Large amounts of taxation, again referred to as

subsidy, have been handed over to various offshore industries. A policy supported by politicians of all parties to provide the country with wind turbines that consistently failed to produce output and solar panels which equally were underwhelmed when the sun failed to break through the clouds.

We are informed that the installation of ground heat pump equipment will save the day. However their energy output begins to fail when temperatures drop to 5 degrees or below.

In areas of the Pennines, prior to and during the industrial revolution, water wheels were used extensively and these were a valuable source of motive power for the textile industry. As with most of the country, windmills were seldom constructed since the country, at that time, suffered from a notable lack of consistent wind.

In a sane world it would be preferable to have policy guided by science, whereas the present situation is that of science guided by policy. These policies are of course backed up by the BBC, academia, and other media whose mandate appears to be to inform the public, particularly schoolchildren, not of how to think, but of what to think. The moulding of public opinion is continuous and any deviation may well result in charges of heresy.

Even if we accept *per se* that reduction in carbon dioxide by some figure, say 50%, possibly going to zero, then there is still a question as to why the reduction to come is from activities based on fossil fuel. Coal fired power stations, petrol and diesel engines and investment in new fields for gas and oil will all be done away with. These are strange targets to pick since they are not, according to reputable sources, the main culprits.

Some years ago whilst preparing presentations for schools to support their science activities, the impact of the built environment on a region was expressed as a temperature gradient from a city center outward to the countryside.

There was a puzzling request to remove this segment from presentations without explanations as to why.

Many years later, when the climate debate was "settled", it was noticed that the evening weather forecast began to occasionally include temperatures in city and countryside in a most matter of fact way. It was almost as though there was an under the radar approach to giving clues out that pointed to weather instability linked to the built environment rather than the preferred script of "global warming".

Also, whilst there is no substantive correlation between atmospheric carbon dioxide concentration and flooding in this country (a correlation expressed as "due to global warming") there is a direct correlation between the over-population of this country and its effect on the built environment.

There are a number of interlinked mechanisms at work. A green field site converted to housing has a micro and macro effect on the environment. The micro effect is to close off the aquifer from rainfall. Water cannot penetrate through brick and concrete and is instead directed via drainage at high velocity to the river or surface storm drains.

On the macro scale the rapid movement through drains collectively overburdens the system and produces storm overflow from the sewage works and flooding from the river.

This imbalance and overload is matched by the above ground temperature effect. On a given day when readings were taken there was an 8 degree Centigrade air temperature difference between the city of London and its outskirts in Epping Forest. Moving from a completely built-up area through garden suburbs to open country showed the correlation between brick, concrete and green field.

Strangely although country and city temperature differences are now mentioned on evening weather bulletins, the scale of temperature rise is not commented on. If the increase in temperature of 8 degrees was given out as a result of the global warming scenario then one could imagine near panic in the media. Yet the unexplained rise between city and countryside remains as a simple comment on the evening news.

There are however at least three factors, an understanding of which adds support to the hypothesis put forward in this book.

These are: Albido

Heat transfer

The fate of rainfall

The albido of a material is a measure of its ability to reflect sunlight. Since the surface heat of this planet is in large part caused by sunshine then the albido indicates whether a material will reflect or absorb – and hence heat up. The scale for this is from zero to 1, with roads it is black tarmac at 0.1, green grass at 0.4 and white paint at 1.0.

The heat imparted to a material is expressed as

sensible heat or latent heat.

The heat arising in roads and building materials is virtually 100% sensible heat and raises the temperature of the material absorbing it. The air temperature above the road or pavement is of course heated by convection from the ground in the same way that a night storage radiator functions.

The latent heat is that required to change the phase or state (e.g. liquid to vapour) of the material absorbing it. Water for example will take sensible heat to raise its temperature to 100 degrees C, with all further heat input being used as latent heat and changing its state from liquid to vapour. The volume of liquid water as it changes to steam increases by 1600 times (the basis of steam engines.) The density of steam is therefore very low compared to water and it rises in the air.

Air moving over water will absorb water vapour and move it up into the atmosphere to cool and fall as rain. This mechanism forms the cycle of heat transfer in the eco-system. Sunlight warms the water and the vapour emitted is picked up by the wind to appear as clouds in the atmosphere. If plants are involved then their considerable pumping action termed 'transpo-evaporation' adds enormously to the movement of water and hence heat into the atmosphere.

Grassland or a planted area will therefore warm or cool depending on the water content of the soil. Water has the highest latent heat capacity of any liquid and is therefore the most useful heat transfer liquid. This process is perhaps best illustrated by comparing the value of latent heat and sensible heat in a saturated soil and a moist soil.

	Latent heat	Sensible heat	Soil
i) Wet soil	1730	-4	-33
ii) Moist soil	940	289	+142
iii) Sand	100	500	+1100

In the first instance of wet soil the soil is cooled as water removes heat from it into the atmosphere. This is reflected in the minus sign recorded for the "-33" figure recorded. In the second, "moist soil", there is insufficient free moisture to evaporate at the soil surface and the soil is heated up generating a plus value. And in the third case considerable heating of the sand takes place since all of the film of moisture coating the sand is free to rapidly move into the atmosphere.

These figures are derived by use of the Penman equation that is used to demonstrate the relationship between solar heat input and surface (soil) temperature. It is quite clear from the use of this equation that the moisture evaporating from a grass field will lead to cooling of the local area. Equally a dry field or surface will heat up.

The moist soil also illustrates how the continuous input of heat (solar) will dry and dessicate the soil. Eventually, with air replacing the water in the soil then the micro-organisms dessicate as well and die off.

The organic carbon in the soil will oxidise to carbon dioxide until only the mineral constituent is left. By itself this material (sand) can be wetted but cannot hold onto water and we have achieved the desert scenario. Without the cooling effect of soil moisture the desert area becomes totally inorganic and the soil cannot sustain life.

In the construction industry materials such as bricks and cement tend to be inorganic and similarly they cannot hold water to any degree. As with a night storage radiator the heat input (solar) falling on the buildings simply acts to raise the temperature. Without water for heat transfer the heat in the city centre continues to build and forms a "heat island".

In the country the water contained in the soil takes the heat and transfers it via water vapour to dissipate into the atmosphere.

The process of evaporation and condensation as rain is one of the fundamental cycles used by nature to balance the planet's thermal inventory. As we expand more and more cities and continue to reduce the greenbelt with housing, we are altering the heat transfer balance and producing instability.

This disparity between greenfield and housing can be further illustrated by figures on the fate of rainfall hitting the ground. In parts of the UK a figure of 1000 millimetres per annum falling as rain would be quite common. This equates to 10,000 tons per hectare of rainfall per annum.

The distribution of this 10,000 tons is as follows:

	Per hectare grass land	**Per hectare housing**
Evaporation:	4000 tons (slow)	3000 tons (fast)
Down to aquifer:	5000 tons	1500 tons
Run off:	1000 tons (slow)	5500 tons (fast)

Thus every area of greenfield converted to housing is generating over the year three thousand tons of fast evaporative contribution to humidity and a net four thousand five hundred tons of excess rapid water run off which, particularly if construction is in a flood plain as is becoming common, will overwhelm local drainage systems.

Thus the humidity in UK cities continues to increase year on year. This is a function of fast evaporation from relatively hot pavements and roofs. The countryside enjoys a more balanced release by evaporation since soil tends to hold water and release heat more slowly than puddles of surface water.

The aquifers in parts of the country are tending to fall since their replenishment by rainfall is inhibited. The reduction in aquifer recharge is not only a function of surface capping by houses and roads, but is in large part due to the impact of modern farming. As mentioned previously, the modern practice tends to speed up the loss of topsoil and to close off the aquifer from rainfall.

The drive to construct more houses on green belt should therefore be seriously questioned and only brown field land should be considered. The impact of any project that would cover the soil surface should have the impact on the underlying aquifer and surrounding water courses as the primary determinant.

It should be born in mind that the rivers and waterways of this country established themselves in a stable situation whereby rainfall on grassland drained into them only slowly. Thus the river channels were deep enough to take the input from rain and deliver it to the estuaries and out to sea without flooding the land in between.

Examining the figures shows that the traditional slow run off of rainfall from an agricultural or greenfield area has now grown to a rapid run off from urbanized development.

Additionally the loss of capacity of the rivers brought on by the displacement of soil from farms into the river bed has put catchments into crisis.

Lack of dredging and urban development has produced regular flooding which is totally unrelated to our alleged "global warming" crisis.

To add insult to injury, or rather injury to insult, the drainage system of this country conveys both surface rainwater and sewage together to the treatment works. Actually the sewage works can seldom cope with the volumes and storm overflows are continuously discharging untreated sewage into the flood waters that surround many new build areas.

For Temperature, CO_2 and Agricultural Area illustrations, see overpage.

Temperature and CO_2 for the last 1,000 years

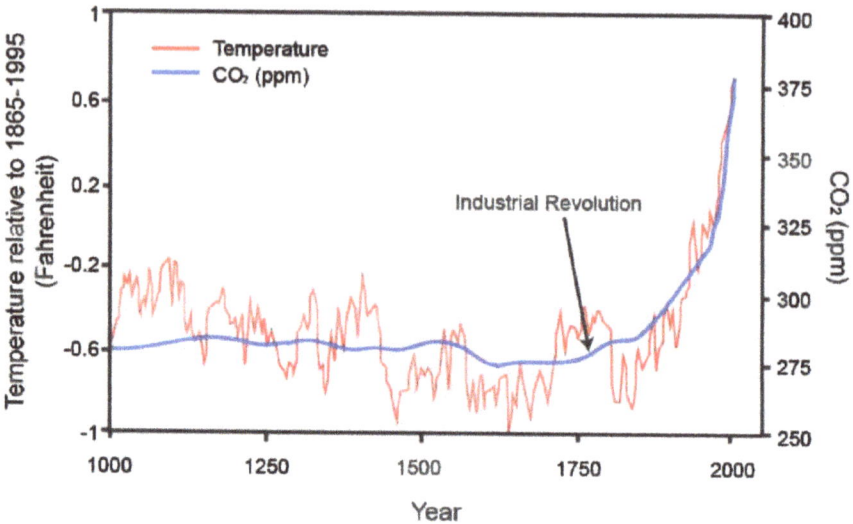

Agricultural Area over the long-term: 1000–2016

Total areal land use for agriculture measured as the combination of land for arable farming (cropland) and grazing in hectares.

If reparation for carbon dioxide emission is due then who should pay? The industrial west or agricultural south and east?

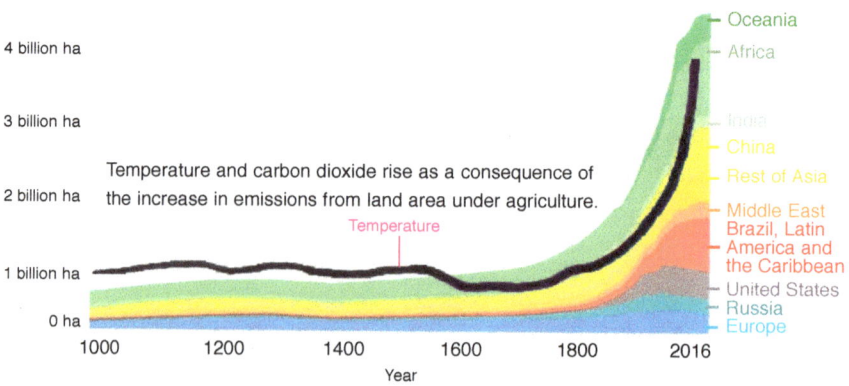

Source: History Database of the Global Environment (2017)   OurWorldInData.org/land-use-CCBY

Chapter 2

Population

Previous desertification of large areas of the planet are evident when studying any picture of the earth.

Documentaries show many archeological expeditions digging in the sand to recover buildings and artifacts of the lost civilizations.

The great agricultural civilizations of the Indus valley were eventually forced to migrate as food productivity fell and drought set in. These civilizations were told by their priests that the gods were angry with them. Unable to comprehend that the continuous extraction of carbon, nitrogen, phosphorous and other elements to grow their basic foodstuffs was also destroying the structure of the soil and particularly its ability to hold water.

Documentaries occasionally say that a civilization was overtaken by drought and they had to move on and abandon their cities. What should be said is that in their lack of knowledge of soil management and regeneration the disaster was self inflicted.

We however have no such excuse. The present religion of global warming espoused by the media, academia and promoted in schools as a matter of scientific fact is a deliberate manipulation of the truth.

The equations and knowledge used to shape this book and examine water loss and soil temperatures were evident in the 60s and 70s, but have somehow been deleted from our vocabularies. The older generation brought up with GCE chemistry and physics would have called the academics and media to account for their fake paradigm.

So let us return to a few basics. The biosphere that covers large areas of our planet was formed millions of years ago. Fungi and lichen that dissolve minerals from the basic rocks of the planet working together with algae, which have photosynthetic ability and feed off carbon dioxide, began to spread and produce the organic and mineral composition of the basic soils.

With complexity increasing a rich organic biosphere of algae, fungi and bacteria began to spread and stabilize. This biosphere was also able to retain water. The next innovation supplied by nature was the growth of plant life.

The emergence of plants with their ability to pump water into the atmosphere enabled the water cycle to be established, and with the plants' ability to access solar radiation on a very large scale, the nutrient, water and thermal cycles on the planet were stabilized.

Thus the carbon from carbon dioxide was sequestered into the soil through algae and bacteria, nitrogen was sequestered by bacteria and, in addition, lightning storms produced the nitrates which act as fertilizer. Phosphorous was taken from the rocks, again by bacteria. Finally the key ingredient of the 'soup' was water, acting as the heat exchange medium and maintaining temperatures within boundaries for human and animal life to progress.

Any damage to this 'radiator' that allows water to be lost and the engine overheats and begins to struggle causing irreparable damage to the vehicle.

Similarly without repairing the radiator it cannot hold water and therefore adding water will simply produce a flood without benefit.

Thus the flooding scenario put forward as arising from carbon dioxide induced global warming is incorrect in a number of ways. Firstly, the loss of humic soil carbon as a result of modern farming practice renders the soil incapable of holding water.

Without water as the heat exchange medium the transfer of heat out of the ground into the upper atmosphere where the heat is released and clouds form, cannot take place and the resulting cooling effect cannot happen. Instead heat islands occur and produce the convective high velocity rainfall that further damages the soil.

Secondly the destruction of soil structure and resultant transfer of slurry into the rivers and water courses reduces their capacity to carry the water flow. The flow of rain is experienced as a short duration, high velocity event that overwhelms the river channel.

Thirdly we should consider in addition to the physical impacts, construction, farming and so on that there is the matter of political policy impact on society that sets the framework and has become a major determinant.

In Victorian times the unregulated expansion of industry was seen as a good thing. This brought both wealth and poverty. In parallel, however, was the undertaking of public works for public good. The provision of drinking water, a sewer system to improve public health, rail transport and so on.

Catchment engineering in Victorian times constructed, in Yorkshire alone, over 120 reservoirs and dams. The largest of these, the Grimwith, can hold almost 22 million tons.

These dams held the water for the woollen textile industry and public drinking water. The water demand for the industry was huge with 1 ton of wool requiring 500 tons of water to process. With the collapse of the textile industry the water demand fell.

However the government website* shows that the population increase that occurred between 1980 and 2021 as a result of government policy was of the order of 12.3 million, almost entirely by immigration.

This figure is more than the total population for Scotland (whose population fell in the same period), Wales and Northern Ireland that together are 5.5, 3.2 and 1.2 million, or 9.9 million in total.

Although immigration is always quoted in UK totals this is deliberately deceptive since it is the cities and catchments of England that have born the brunt.

Thus the reservoir capacity constructed for industry served instead to supply water to this disproportionate increase in population and this brings us to perhaps the biggest impact of all on our catchments, the physical impact on demand and services that mass migration brings.

The big surge in population occurred during the years between 1980 and 2021 particularly when the

---

*https://www.ons.gov.uk/peoplepopulationandcommunity/populationand migration/populationprojections

"primary Purpose" rule that had maintained a break on immigration was abolished and it was mandated that effectively any person arriving in the UK was entitled to benefits payments*. The impact was dramatic, indeed the consequences are still with us today.

At the same time the de-industrialisation of the UK left the country with six out of ten of the most deprived areas of Europe and with a political class uninterested in the problems of these derelict areas.

However, to counter the possibility of the English electorate turning away from the politicians who were culpable it was decided to import a new electorate in to the key areas. In time this new electorate will outnumber the English

A significant and unplanned factor in this dramatic population rise however was the impact on water demand. Demand is usually calculated as 200L P.E./day which in this case translates as $12.3 \times 0.2 = 2.46$ million tons per day. Whilst the supply side of drinking water was taken care of the reverse side of the coin, an equal volume of sewage output, was not.

The calculation of sewage volume is, by the same standard, factors of "Personal Equivalent." This, as is indicated, is the volume of sewage per person per day requiring to be treated. For an additional population in excess of 12 million the figure becomes approximately 2.5 million tons per day, a large proportion of which is now passed through into the rivers as "storm overflow." Thus the factors described earlier together with the storm overflow situation equates to ever growing numbers of

Broken Vows – Tony Blair, The Tragedy of Power: Tom Bower, 2016

flooded homes and businesses.

Strangely one seldom hears of flooding in Scotland, Wales and Northern Ireland.

Consulting the various maps offered by the authorities which show present and predicted flood zones for the UK, one could be forgiven for thinking that the problem was being shared in much the same way that Chancellor Merkel proposed for all of the EU countries to the occupants of the Middle East to re-home throughout EU.

Closer examination shows however that the maps for England give the majority of flooding as inland around the densely populated areas and the flooding is generally surface water contaminated with sewage. The maps for Scotland are largely for coastal areas being flooded by the rise in sea levels arising from global warming. Inland groundwater flooding does not appear to be a problem.

If you know your population equivalents then you can forecast the number of homes/catchments that will be flooded. This figure is presently quoted in order to prepare the public as 3.5 million homes by the year 2050, none of them however will be the victims of global warming.

Put quite simply, the loss of greenbelt to housing and development, the loss of topsoil into the rivers, which are largely undredged, the inability of rain to penetrate into the ground and the additional sewage, which under wet weather conditions is allowed to pass through the sewage works untreated, are all conditions that have negatively impacted catchment (mis)management.

We can already see a vision of the end game that will come about if the real cause of weather instability is not tackled.

Our Half Hydrological Cycle

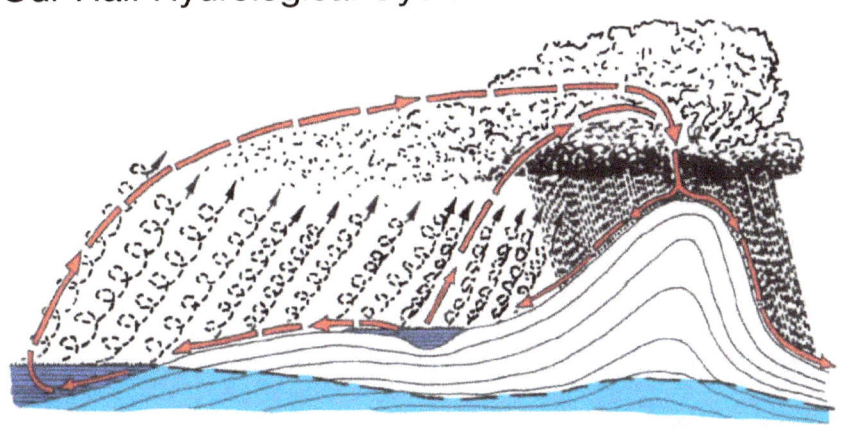

After Scauberger, Kravcik and others

Disrupted ecology and degraded landscapes (Carbon lost): causes rainfall 'extremes', rapid evapo-transpiration, flooding, poor infiltration to (falling) water table and drought.

Our 'Full' Hydrological Cycle

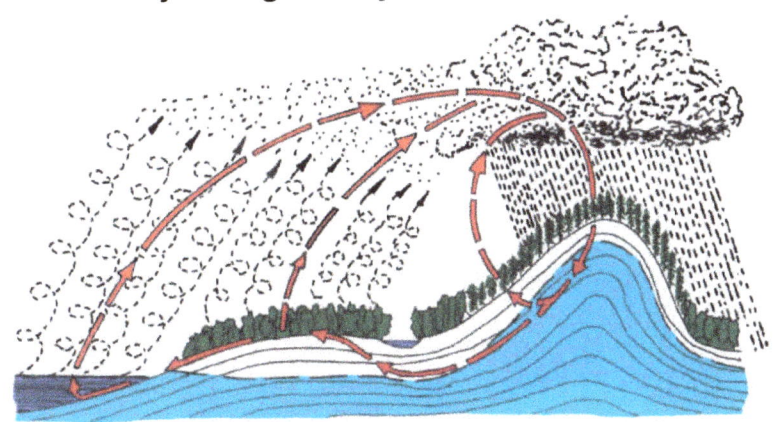

After Scauberger, Kravcik and others

Functioning ecology: Balanced rainfall steady evapo-tanspiration, good infiltration to water table.

Chapter 3

Fossil Fuels

Analysis of the IPCC table of emissions lists the activities and associated carbon dioxide outputs that allegedly are producing climate change.

Hand in hand with this is the focus on ending the use of fossil fuels and replacing the present energy supply required with some form of "sustainable energy." This we are told will prevent the catastrophic rise in temperatures that could end with a run-away heating of the planet's ecosystem.

Whilst there is no doubt that the huge expansion of global industrial and agricultural output was due to "big oil" the opposite opinion that to remove big oil (and coal) will lead to a sustainable future is, from the evidence, extremely tenuous.

The removal of fossil fuel (oil/coal) from societal use will have negligible impact on the climate change scenarios put forward. Indeed the change to "sustainable" energy, wind and solar, will soon change the word to "unreliables" in the protestors' phrase book. It is more than likely however that nuclear will come to the rescue and we will at last be living in the clean air resulting from all electric vehicles.

Unfortunately, since fossil fuels account for only 13% of the carbon dioxide output (agriculture 23%) the effect of this colossal economic dislocation (at a price of several trillion pounds according to the treasury) will have zero impact on the runaway scenario being put forward.

And although "clean air" may see reductions in the sulphur and nitrogen dioxide contaminants resulting from vehicle exhaust emissions the air pollution levels will be more than replenished by those toxic elements coming from wildfire/forest burning and the dust particulates as the fine sand produced by the destruction of soil carbon cover the sky and occlude the sun.

Reading or watching the news on any given day in 2022 produces the following partial, list of impacts:

In America during April (2022) some 20 wild fires were burning in New Mexico, with one of 84 square miles in area. Arizona was suffering a 175 mile square forest fire resulting from a 20-year mega drought with severe aquifer depletion. At the same time, 80% of the American west is facing severe drought, and so far this year US forest fires have burned twice as much area as those in 2021 and 70% more than the 10 year average.

Indeed a column of fire rose into the sky and when it collapsed burning embers and sparks began to fall onto Los Angeles. All land in the region is drying out and the end scenario will be the dry desert situation devoid of both aquifer and surface water.

Previous generations remember this "dustbowl" effect with bad farming practice leading to the soil turning to dust as all organic carbon was lost and with it the water holding capacity. Failure to learn the lesson and overuse of aquifer water has brought the same disaster on today's agriculturalists in California.

Elsewhere, according to the BBC, Sky *et al*, Pakistan has seen the hottest March in over 60 years with neighboring parts of India suffering a similar heat wave. Temperatures were over 6 to 8 degrees higher than normal with almost 1 billion people at risk from excessive heat. Authorities are preparing for floods from the thousands of lakes formed in the Himalayas from the melting glaciers. Over 30 of the largest of these lakes are at risk of sudden discharge to produce serious flooding.

In the Caribbean region a vast area of foul-smelling seaweed continues to grow and devastate the coastal areas upon which tourism depends. The culprit in this case appears to be the discharge from the Amazon river basin of water rich in nutrients and sediments from the new agricultural activities in the deforested areas. Adding to that is the discharge of nutrient water from the Congo basin. Both of these flows merge, collide with each other and surface in the Caribbean area of the ocean to produce the seaweed.

To summarise, calling for the elimination of fossil fuels, coal and oil will remove 13 billion tons of carbon dioxide per annum from the inventory. Agriculture, as practiced, accounts for 23 billion tons and together these account for less than 40% of the total. So why remove fossil fuel and produce extreme dislocation of the world's economy?

In particular the proposition that fossil fuel burning is the root cause of climate change is in itself not proven being simply a majority vote backed-up by a correlation between carbon dioxide and temperature. Correlation however is not causation. Indeed the correlation between Carbon Dioxide, temperature and the increase in the land being dedicated to agriculture shows a better correlation. However, in all cases we are seeing an effect not a cause.

Chapter 4

Ecosystems

The eco-systems of the earth have been assembled over millions of years. The algae and fungi coming together and spreading over the cooling rocks, soaking up carbon dioxide and nitrogen from the air and rain. The water cascading down the mountains displacing rocks and grinding the hard rock into small particles. Lichen extracting metals and phosphorous from the slurry.

Carbon, nitrogen, phosphorous, and metals coming together to form plants. Plants that spread rapidly using the power of photosynthesis to create a vibrant world with a multiplicity of forest and plant life. All of these dependent on water to connect and cool the eco-systems and maintain the thermal balance of the planet. The reduction in carbon dioxide produced by plant growth and the cooling produced by evapo-transpiration produced a planet capable of supporting animal and human life.

The direction of complexity was always upwards with

each part fitting into a symbiotic interdependence. Sadly, mankind looks not at interdependence but exploitation. The complex organic soil developed by nature produced excellent food full of all the nutrients required for the good health of humans.

How much better however to mine phosphorous from the ground, combine nitrogen with natural gas and put the resulting product into bags and create a profitable market that generates dependency. And so the degradation of soil accelerated.

Excess phosphorous and nitrogen in various manufactured forms now pollute the aquifers of the UK, making many of them unusable for drinking water. Soil, degraded and broken down, saturated with phosphate fills our rivers and is carried into our estuaries. In many parts of the world lakes and rivers are being choked with algal blooms.

Carbon, together with its cohorts phosphorous and nitrogen, form the basis of life of the eco-systems, and their removal from the natural soil cycle and replacement with concentrated fertilizer is bringing about a transition back to a time when the earth was a desert devoid of life. The large excess of fertilizer serving only to pollute aquifers, rivers and seas.

Water, sequestered in soil for so long is being released from the land and has produced the majority of sea level rise. High velocity rainfall instead of being contained in the land for use by plants runs off rapidly to produce flooding.

The land left behind is dry and dangerous. With no cooling effect available sensible heat produces substantial rise in temperatures and the volatile oils and resins in trees and plants form a ready source of fuel to be ignited by the tiniest spark.

In the recent induced panic over the breaking of all temperature records when the fires of hell reached the surface of our world a simple experiment demonstrated the point.

Temperatures taken at a location in Leeds (a major heat island) gave readings in line with the weather forecast, around 40°C, this however was the air temperature, the ground temperature (concrete) recorded at a peak of 56.3°C around mid afternoon. This indicates quite clearly the impact of sensible heat build up in cities and also shows the temperature direction that further development with concrete and bricks will take in the environment.

One might add that air conditioning (AC) will become a requirement for future housing as it is in offices. Careful consideration however points to the fact that AC, which takes warm air from inside a building and dumps it outside will add to the external heat load and hence temperature of the Urban Heat Islands (UHI) that cities are becoming.

Aquifers too are pumped dry or are polluted and dangerous to drink. Constantly however we are told that hard engineering will solve the problem. It is true that the hard engineering of concrete pipes will carry away the flood water to a different part of the catchment, but the primary purpose of water is not to submerge whole areas with storm sewage that carries a high viral load, it is to use the water for the people and agriculture.

It is time perhaps to consider other forms of engineering provided by planet earth free of charge. In the next section of the book we can examine systems of natural earth engineering that deal effectively with water pollution of all types and do so whilst at the same time sequestering Carbon, Phosphorous and Nitrogen.

Consequences

Loss of carbon from agricultural soil – recorded as 5 tons per hectare per annum owing to modern agricultural methods.

The resulting inability of "soil" to hold (rain)water and its conversion into an unstructured slurry that washes into rivers producing blockage, overtopping and flood. Leaving behind a panned impermeable surface.

The rapid transfer of water from developed land at a rate too intense for drainage to cope together with the additional overload from sewage as a result of unsustainable population / infrastructure relationships.

The loss of 4 key essential components from the ecosystem whose presence nourishes and protects all life on the planet, i.e. water, carbon, nitrogen and phosphorous.

A failure to comprehend the symbiotic relationship between these components.

The failure to prevent river/estuarial discharge of nutrient rich water from the land into the ocean.

The failure to prevent agricultural run-off of pesticides, phosphorous and ammonia (nitrogen) from polluting the aquifers and rendering the aquifer water unsuitable for drinking or irrigation.

Examination of these consequences in turn serves to illustrate the scale of the problem, a problem that is being obscured and camouflaged by the emphasis on fossil fuel eradication.

*

Chapter 5

Water & Carbon

These two together underpin every activity. Without water all life forms dessicate and die. In moist soil the bacteria carry out soil respiration and the gases carbon dioxide and oxygen are continuously cycling in much the same way that human respiration continues.

Carbon, as constituted in plant roots and the multiplicity of soil organisms is constantly alive and moving internally within the soil. Water is the soil solution that maintains nutrient levels in balance and importantly enables the transfer of heat or cold depending on external stimuli.

A deeply ploughed soil has regions consisting of aerobic (oxygen rich) and anaerobic (oxygen deficient.) As the soil depth increases the bacterial situation becomes anaerobic. Within this realm however there are many critical reactions that all play their part in cycling and recycling nutrients within the soil solution.

However flooding of soil for long periods produces a completely anaerobic system that may generate toxic waste products, that poison other micro-organisms within the soil.

Terrestrial systems therefore require alternate rain and dry periods to allow the soil areas to recharge and then drain.

Closer to the surface the soil becomes aerobic and the bacteria in this area tend to be the ones synthesizing plant food, which is then transferred into the root mass of the plants. This area is more porous and open to atmospheric oxygen.

As water is removed however the soil processes begin to falter. If water is removed altogether then the soil processes cease and the life-forms within the soil become dessicated and die. At this point the atmospheric oxygen is still penetrating and those materials with a carbon framework will oxidise to carbon dioxide. With the emission of all carbon as carbon dioxide the ability of soil to rehydrate is compromised. Rainwater will still penetrate but there is no mechanism to bind it into the soil.

At this point the heat exchange mechanism also fails. Heat, from sunshine, is taken as sensible heat by the soil. Any residual water uses this heat to evaporate off – the radiator has been punctured and the water has drained away – all that remains is desert sand, dry wood and other dessicated, carbon based material. As the incoming sunshine continues to heat the ground and the dry trees, the resins and oils in the wood become dangerously close to their flash/ignition point.

The hypothesis put forward in this book therefore is that it is the interruption of the hydrogeological cycle of the planet globally and locally that is causing the present climate change scenario.

Heating a kettle of water will eventually vaporise the contents and the lite vapour boiled off will rise into the atmosphere and cool, surrendering its heat into the atmosphere to cool. This is exactly the mechanism that takes water vapour from the ground and moves it upwards to condense as clouds. The heat supply in this instance however is the ground itself that is thereby cooled.

For decades now modern agriculture has been lightly ploughing the soil, opening the structure and oxidizing the soil's humic carbon to carbon dioxide. The amount of carbon dioxide released from this source is far greater than that from the use of fossil fuels, and the demonizing of fossil fuel is a deliberate distortion of the reality.

The significance of carbon dioxide emission from soil is not that it produces a greenhouse warming effect, but that the heat exchange capability of the soil is eliminated.

The outer skin, or echo system of the planet, a living symbiotic layer upon which all life eventually depends, has caught a fever. When the fever has eventually and literally burned itself out then all that will be left is a lifeless inorganic skeleton of sand containing the bones of another lost civilization.

Globally the number of countries suffering the drought/flood scenario is increasing and the picture is almost always the same, with the media showing a sand blown desert with the red pale brown color of unstructured decarbonized soil.

In the UK and Northern Europe generally the green cover is still evident and the UK, with its maritime climate, is unlikely to turn to desert in the near future, although Norfolk and other agricultural areas are prone to the flood/drought scenario typical of soils that are unable to store water.

The climate change scenario is now being shoehorned

into every disaster event. The spin doctors know that constant repetition is effective and drowns out any contrary analysis of the latest calamity.

Regular flooding of parts of the "UK" are all "global warming" effects and nothing at all to do with being one of the world's most overcrowded countries, with sewage systems that make little attempt to capture and treat sewage, but opt instead for dilute and disperse through storm overflow.

Greenbelts and floodplains are now prime targets for house building as developers see this as a cheaper option than remediating contaminated land. Additionally a greenfield is perceived as more aesthetically desirable than an industrial area. Of course the hidden cost to the environment by isolating the aquifer and the incremental damage to the water cycle is not costed in.

Job creation is often used as the lever to free up development but the jobs tend to be transient and disappear from the area once development is completed. We are promised that the flooding problems that many catchments in England suffer from will be dealt with by a raft of new engineering solutions such as "resilience," or "sustainable urban drainage." These phrases are as useless as the years of promises to curtail population increase.

A true engineering solution could do no better than remembering that it is difficult to see the wood for the trees.

Historical data is clear in that there has been a large reduction in tree cover over the last thousand years in the UK. Some recovery has occurred in the last 300 years and attempts are being made to accelerate the effort as part of carbon sequestration.

The rational and basis for this is the fact that trees

require carbon to grow thereby reducing atmospheric carbon dioxide. This fact is self evident of course but it also highlights the failure to assess the true nature and potential of trees in helping man combat the climate instability being experienced at present.

People are beginning to also realize that clearing the 'rainforests' of S. America in order to transfer agriculture from Europe is perhaps an unwise thing to do. The unease is perhaps to do with the title: "rainforest"

In other words there appears to be connection with the worlds water cycle and trees. In reality over 60% of terrestrial rainfall is produced by trees and the soil they grow in. The remainder arises from warm air moving over the seas.

It is unequivocal that trees play a pivotal role in the functioning of water and energy cycles and thereby in temperature regulation. And that the conversion of surface heat, sensible heat, to latent heat, through water evaporation leads to cooling of the air.

A compilation of research papers in the US has gathered together some forty papers on trees in many aspects such as trees in the city, in arid conditions, types of trees and so on.

In each case there is an attempt to establish the impact and opportunities that trees present to persons involved in making life better in cities by reducing the temperatures within the UHI (Urban Heat Island).

The driver was the sad fact that in the US the number one cause of death as of 2017 was the impact of heatwaves.

The initial observation was that trees block solar radiation from hitting the ground by up to 90% and the shading effect of trees alters the albido, substantially producing up to a 20°C reduction in surface temperature.

This of course depends on the type of tree and its canopy. For large canopy trees there appears to be a double benefit in that day time benefit of cooling is matched by a slight rise in night time temperatures by trapping heat in their canopy and the trees are acting as a form of air conditioning.

This becomes significant when one realizes that globally cities consume 75% of the world's energy and account for 70% of Carbon Dioxide emissions although it is not clear whether this is a collective model and includes all inputs to city life.

A second and major effect observed was the fact that the energy used by trees to evaporate water (evapotranspiration) the latent heat, is drawn from the sensible heat stored in the ground. The outcome of this interaction is that the surface heat is reduced and the ground temperature drops followed by a reduction in air temperature.

Measurements in Los Angeles show that trees moved more than 135,000m cubed of water per day to the atmosphere, this in turn pulls sensible heat from the ground and drops the air temperature by up to 8°C in the local and regional area.

Strangely, the policy of planting more trees still seems to be based on the adsorption of carbon rather their cooling capability. Additionally trees appear to be planted as monoculture which inhibits biodiversity and the run off from pine plantation is acidic giving increased cost and energy use for the water companies.

In engineering terms the use of electromechanical air con which cools a limited internal area but dumps heat externally thus adding to the city temperature and also is using electricity whose generation is releasing more carbon dioxide emissions is preferable to using trees, which are solar powered, cool the ground and air temperatures and absorb rather than emit CO_2. There is room here for more than a little lateral thinking.

A surface temperature reduction of 20°C and an air temp reduction of 8°C over a whole region and the problem then becomes how to keep feeding water to the trees. In other words how do we return water to the land in order for the trees to keep the ecosystem cool.

Chapter 6

Policies

It is time to consider a different approach involving catching and storing water on the land.

The four main components of the problem are, as stated previously, water, the soil solution within which all activities take place, and carbon, nitrogen and phosphorous. Together these three are the most critical for all life processes.

The climate crisis produced by loss of soil carbon manifests as erratic weather, flood and drought and large scale forest fires. They are all physical manifestations of eco systems in trouble.

The crisis generated by the imbalance of nitrogen and phosphorous are not at the moment as visible or dramatic but the warning signals are there.

In the UK many aquifers and surface waters are polluted and undrinkable. Not just from nitrates resulting from agriculture but also phosphates from sewage.

In Malaysia it is a common occurrence for ammoniacal

run off from chicken farms to knock out drinking water supplies, and in Sri Lanka food shortages are exacerbated by a sudden drop in fertilizer application. In this case it was not realized that using fertilizer to grow crops is equivalent to dosing drugs to humans. When the supply stops, then withdrawal symptoms appear and the system crashes. A good soil suitable for farming can be re-established, but requires a minimum of two years to become functional.

Just as there is no doubt that the mechanical methodologies of industrial agriculture are causing the emissions of more carbon dioxide than the burning of fossil fuels, so the chemical fertilizer activity is doing severe damage to aquifers, surface waters, river basins and the sea.

This is not to say however that the damage cannot be mitigated but a requirement for such mitigation is that politicians understand the problem and in doing so they then act on behalf of the people.

Sadly the railroading of public opinion by sections of the media without any clear scientific basis and a commitment to a 'net zero' without a properly costed strategy leaves too many people feeling that this is not happening.

However, for people to be surprised by the proliferation and unanimity of academic opinion in favour of carbon dioxide induced climate change and "net zero" policies simply demonstrates a lack of knowledge of how the present system of governance works.

The privatization of higher education ensured that "he who pays the piper calls the tune" and this factor dramatically alters the process of academic research.

In the 50s, 60s and 70s a majority of school/university leavers were employed in industry. A great deal of R & D was carried out in industry by, for example, ICI and the methodology and rigour of industry and science was applied to research projects. Science tended to advise and direct policy for government.

With the growth of trans-national groups, such as the EU, funding and control of R & D moved away from industry to one of the many "Directorates" in Brussels. From there it was fed back to academia and industry provided the programs presented were acceptable to the Directorate.

In the same period the manufacturing industry of the UK was exported to Europe leaving the UK with essentially de-industrialised areas of former coal, steel, shipbuilding and chemicals, areas now referred to as the red or blue wall.

The funding by industry for R & D was gradually replaced by funding from a continuous succession of research topics stipulated by those who were major influencers within the EU. Essentially academia became dependent upon government not industry for their funding. This happened almost imperceptibly and cartels were formed between academia and civil government, centralized and directed by Brussels.

The essential difference between the two systems is of course that industry requires a justification for its investment, and a new technology must be proven to be superior to existing technologies for it to succeed in the market place.

For academia and bureaucracy, the objective is very often the process of research itself irrespective of outcome, and additionally, technologies that are inferior to those existing can be justified without due diligence and usually with large subsidies from the population.

Thus the centralized bureaucracy employs policy to direct science. Therefore, vast subsidies (tax) for e.g. wind farms and solar projects, which are stunningly ineffective and intermittent in their energy output, are paid. This can also mean that cartels are enabled to capture and copy technologies that strictly speaking they have no real entitlement to.

A classic example of this activity is the implementation of a technology known to almost everybody as reed beds. Around 40 years ago this technology, based on soil science and microbiology, was demonstrated to be effective in the Water industry, challenging existing technologies for cost effectiveness in waste water treatment.

Beginning with the treatment of sewage by Prof. Rheinhold Kickuth in Germany and developed into toxic waste water treatment for industry by the Root Zone UK group (RZG), the technique, called at the time Root Zone method, was showing excellent application to all types of waste water.

At that time, the UK Water industry had formed a central research group to explore new technology. In order to sell into the industry, new technology had to be assessed by this central group and "passed" for application within the UK water companies.

Root Zone UK (no longer trading) therefore presented information to and organized an inspection visit for all UK Water authorities to various Root Zone systems. The report from this visit drawn up and approved by all UK delegates was extremely favourable, and recommended adoption of the process into the UK sewage sector. What wasn't realized at the time by RZG was that the water industry was to be privatized and it wanted to extend its portfolio of treatments.

The timeline for reedbed/wetland development was therefore approximately as follows. The central industry group applied for and received funding from the EU to centralize and develop the concept of reed beds demonstrated by RZG and to prepare guidance and design protocols.

Funding for research by academics and organisations linked to the research group was extensive and resulted, in the following years, in a number of conferences initially in the UK but eventually on a world wide basis through academia and government funding.

Articles and journals generated by this activity had a strange lack of acknowledgement of RZG and failed to mention any of the systems that had been proven and operational for many years.

Of great importance to the concept was the fact that the soil, plants and microbiology were fundamental to the dynamics and operation to the system.

Sadly, this knowledge was discarded in favour of simple gravel plant systems which are much easier to build and are totally lacking in the symbiotic grouping of soil/plants/bacteria. Forty years of literature, however, and extremely expensive research funded by public money demonstrates that these systems are completely ineffective for the majority of applications.

The academic stranglehold of Water industry/ government has prevented the adoption of the technology in the sector where it is most needed – the Water industry.

We have therefore a precedent for the 'emperors new clothes'. A large organization with representation at all levels throughout the world and with academia never questioning the obvious failure of the existing 'fake' technology.

The systems built for industry by RZG and reported later in this book achieved huge advances in meeting the environmental problems currently surfacing throughout the world. These systems clearly demonstrate Carbon sequestration, the removal of Nitrogen in agricultural runoff and Phosphorous entrapment on a large scale.

A statement from a Water Research Center report almost forty years ago states:

"Phosphorous removal occurs by chelation and precipitation in the clay soil."

Following recent technical presentations to various sections of government departments by former RZG personnel, it is understood that once again grants have been given to academics to study Phosphorous removal.

History appears to be repeating itself and no doubt another forty years of wasted effort will produce "experts" to be wheeled out by government following a further vast waste of public money.

In the meantime existing systems continue to sequester Phosphorous from polluted water and remove it from the environment.

For reed beds substitute 'climate change', and it reveals the same process.

*

Chapter 7

Dust to Dust

In closing the path to soil/plant based treatment systems a major step in understanding and repairing our damaged ecosystems was closed off. There is a fundamental importance to the soil/clay/humic structure in terms of chemical and biological activity.

Additionally it can also be demonstrated that plants function on the mechanical level as a means of maintaining a porous soil structure that stabilizes the pathways and hydraulic conductivity of the soil.

The establishment of plant roots within a soil structure (to create a "rootzone" or "rhizosphere") also further enhances the inherent chemical and biological activity of the ecosystem.

Water, oxygen and nutrients moving through the porous soil provide a living and breathing ecosystem which is fundamental to the survival of all life on the planet.

There exists in the literature no exact classification of

bacterial types in terms of complexity. One can say however, that taking nutrient requirement as a measure of this complexity, approximately seven classes of bacteria can be determined.

1. The simplest, which require glucose/nitrate/mineral salt.
2. Above, plus a minimum of ten amino acids.
3. Those requiring: cysteine and growth factors.
4. Those requiring: amino-acids and growth factors.
5. Those requiring: yeast extract.
6. Those requiring: soil extract.
7. Those requiring: both yeast and soil extract.

Although this list is very simplistic, it does provide a useful general guide since soils of low fertility tend to have colonization of the lower member species, whilst a fertile soil will tend towards the more complex types.

A well rooted high nutrient soil will contain, in addition to types 5 and 6, types 2, 3 and 4 generally associated with the plant root system.

One has to be cautious however in what one classes as nutrient. A deep well rooted soil will 'fix' its own carbon and nitrogen from the atmosphere through bacteria and specialized plants. Plants and bacteria will also extract trace elements and phosphorous from the soil in which they grow.

As the plants develop and grow their nutrient, demand increases and this is matched by an increase in bacterial and photosynthetic activity.

The older system of farming, employing horses or oxen and deep ploughing, broke up the soil surface and drove their manure into the soil to be recycled as nutrient carbon, nitrogen and soil fibre.

Rotating fields around cattle, crops and vegetables maintains the structural and mineral properties of the soil. In this system, now known as 'regenerative farming', the use of cattle becomes imperative and provides a natural food cycle of beef and dairy protein together with vegetables and crops that between them contain a full suite of nutrients essential for healthy life.

The idea that we can live a full and healthy life on a plant based diet alone is simply further evidence of how far we are away from our understanding of mankind and his relationship with the ecosystem.

Fortunately, this type of rotational or regenerative farming is now making a welcome return and is available as the 'Savoury System' among others.

From the 1950s onwards, however, the direction was towards fertilizer dosing. This is equivalent to the use of steroids in sport in that there is a 'bulking up' effect, but nutritional value is compromised. As a result the nutrient status of vegetables and fruit is very poor when compared to similar products from the pre 1940s for example.*

The impact of high nutrient doses shows itself in a number of ways. The usual fertilizer would be expressed as NPK or nitrogen, phosphorous and potassium. The impact of an excess of these in the soil solution is to ramp up the bacteria tasked with the conversion of them into species to be used as food for the plant.

The result, however, is to suppress other key biological processes such as the carbon cycle that are 'crowded out' and as a result the plants become victim to disease and parasites which then require more chemicals

---

There will be a follow up article on the subject of the nutritional value of food and its impact on physical and mental health.

to protect them.

Measurements of photosynthesis show that a plant fed with fertilizer will reduce photosynthetic activity since the cycles that produced food internally are now replaced by external supplies. However, remove this external support and the plant system will collapse. Rebooting the system can be achieved, but is a difficult and time consuming process.

Finally, the excess NPK usually drains off into the aquifer, rivers and all other parts of the water catchment where it is detrimental to the environment.

An additional point often overlooked however is that the soil system is in fact an extremely complex bio-reactor. The expression 'dust to dust' acknowledges the fact that animal and humankind will, when interred in the soil, be subject to complete degradation of all organic components, which will then be recycled to take part once again in the life processes of the planet.

This demonstrates that soil, when treated properly, can break down unwanted material, such as sewage and cattle slurry, and reprocess and reuse the basic elements obtained back into the life cycle. The process of breaking down material with organisms extracted from soil is now known as bioremediation, whilst soil organisms when used in conjunction with plants is termed phyto-bioremediaton.

In old terminology the treatment of sewage by discharge over pastures was known as the land treatment method.

The land treatment method relied upon the shallow surface soil and the oxygen requiring bacteria in order to function; overflooding of this surface layer would drive out the atmospheric oxygen upon which the bacteria depended and the system would collapse.

The example given here of the Root Zone method at Othfresen, a village of around 4,500, was the first demonstration of a substantial phyto-bioremediation project for sewage. The key difference between this and the land treatment system is that the wetland plants used have adapted to provide oxygen from the air down into the soil and root matrix.

The comparative data is presented in the following table:

	OTHFRESEN	ICI
Water Holding Capacity	T0 500M^3 2Ha T16 1,500M^3 subsurface	T0 600M^3 5Ha T16 10,000M^3 subsurface
Soil Carbon Capture %	1,500,000KG calculated as C.O.D. from sewage TO 4% T16 11%	52,000,000KG calculated as Carbon (from C_6H_5OH) TO 3% T25 50%
Emissions Reduction	16 Years 53,000,000KG CO_2	25 Years 1,839,000,000KG CO_2

At Othresen, the soil placed originally in the bed had a water holding capacity of only 500M^3 at Time Zero. Sixteen years later, the holding capacity was 1,500M^3. The soil carbon content was initially 4%, but after 16 years had reached 11%. The emissions saved by using natural systems was 53,000,000 KG of CO_2.

At ICI, the the water holding capacity of the soil initially was 600M^3. After 3 years, the mature saturation point of 10,000M^3 was achieved. The soil carbon content was an initial 3%, peaking at around 50% after 25 years. The amount of emissions saved by using the soil/bacteria method was a staggering 1,839,000,000KG of CO_2.

This greatly extends the depth of the 'reactor'. In addition to the greater depth is the presence of both aerobic and anaerobic bacteria. This also enhances the power of the system and is a system that it has not as yet been possible to reproduce, even by the most advanced engineering.

At the time the system was built to overcome the lack of water being experienced in the bird sanctuary but the remarkable success achieved, giving better quality water than conventional sewage treatment, aroused much interest.

Conversations at the site revealed two significant factors. Firstly, the construction and operational costs were substantially less than conventional civil/mechanical constructed works and secondly, local civil engineering contractors revealed that they would announce substantial job losses just prior to the next elections if the proposed additional reed beds were constructed.

This threat inhibited deployment of further systems in the region.

By the mid 1980s an encouraging development had taken place. Monitoring of energy use of the reed bed against the equal capacity in terms of Person Equivalent conventional works in the same area showed that energy use and consequent emissions for the reed bed system were negligible.

The system was essentially driven by sunlight except for the pump carrying the sewage to the site. Flow through the system and out into the river was by gravity and the soil/plant bacteria housed in the rootzone carried out the mineralisation of the sewage into humic material and plant food.

The conventional works had high construction and operating costs plus it required power station electricity to operate.

In assessing the impact of this soil based technology we can perhaps now understand why a completely bogus gravel/reedbed system was rolled out throughout the water authorities in the UK, and indeed on platforms throughout the world.

The plants and gravel systems tagged on to conventional sewage works did indeed provide an element of 'greenwash' and a small amount of improvement in the final stage of sewage treatment.

Unfortunately, they also acted as settling ponds for small, but continuous, amounts of sewage solids.

These small amounts of solids built up over the period of use and, in conditions of storm overflow, they tended to discharge these solids into the nearest river.

However, with national and international committee structures fully supported at government level of the gravel reed beds and wetlands (IAWAPAC) and with the control of information and publication of research papers, it was quite desirable to become a member of the club and, for a fee, to become accredited as a supplier of this type of 'emperor's new clothes'.

From a commercial point of view the design, construction and operation of a modern sewage works is an expensive undertaking, but operates in a captive market. The customer, the public, therefore, is faced with a monopoly created by the government.

The consequences of these monopolies are all too apparent. The continuous pollution and destruction of our rivers and beaches by sewage discharge. The storm overflow of sewage containing high viral/bacterial loads acting as vectors for illness.

The dereliction of duty by those responsible for

legislation and policing of the sector which, despite many reports and protests, still acts with virtual immunity from prosecution.

Alongside all of that however is an environmental threat that is coming in under the radar.

Drought

For the summer of 2022, a heatwave followed potentially by violent convective rain storms, is the forecast. With newscasts showing empty reservoirs, dessicated vegetable crops that could not be removed from ground that is too hard to work, and acres of crop stubble that pose a major fire hazard, the country is faced with both a food and water emergency.

Convective, high energy rainfall will of course dislodge topsoil and move it to the river where the reduced capacity will result in overtopping and flood. The soil surface that remains will be 'panned' and impermeable and will inhibit access for further rainfall to the underlying aquifer.

All of this of course due to a 'global warming' scenario that brings forth the same disastrous solutions involving large scale financial subsidy that will move the disaster to another part of the ecosystem that is currently untouched.

In this case we can expect to see applications for new reservoirs in all the beautiful parts of our country that have not been built on, the Downs, the Dales the Weald and anywhere else that is as yet unspoiled.

These reservoirs will then be filled by the use of large desalination units built on equally beautiful parts of our coastline.

There will of course be the usual environmental

assessment whereby wise heads will be paid huge sums to conclude that regrettably, though damaging to the environment, the schemes must proceed in order to provide water security.

One must also remember that for the last accounting year a record of over one million immigrants were allowed into the country. Thank heavens that the politicians promise to "take back control" of immigration was in full process otherwise the housing, energy and water shortages could have been even more serious.

Nevertheless, a short calculation indicates that one million people require 200 litres of water per day each and this amounts to an extra two hundred thousand tons per day of capacity.

Equally this will transform into two hundred thousand tons of sewage per day to be treated, or dumped into and onto our rivers and beaches as is becoming a more usual practice.

Anybody familiar with the Gulf states will know of the practicalities of desalination to provide drinking water.

To begin with, the energy use is intensive and it is preferable to have your own oil field to supply the necessary energy, so perhaps we should open up our North Sea fields again since the chance of Europe supplying the gas or oil is rather remote.

Secondly and more important, the residual output from desalination of seawater is hot salt solution and this when discharged back into the sea will render both the temperature and salinity totally unsuitable for the few remaining fish left over from the common fisheries policy.

However, if we examine the pictures we are shown with the drought report from different areas of the country these demonstrate quite clearly that the real problem is one of unstructured, decarbonized and dessicated soil.

Decarbonized soil cannot, as previously explained, hold water, nor can plant roots survive or grow. Worms and all other soil fauna that act for the soil as a self-repairing mechanism will perish and the soil, without help from man, will cease to be a living organism. It is thankfully, possible to re-carbonise the soil and to return water to the land and use soil as a reservoir. It is also possible to deal with the nitrogen and phosphorous contained in our polluted aquifers.

In the next section of the book the methodology of carbon capture from sewage and effluent treatment will be outlined. In this case, however, the purpose is more to enable the soil to act as a reservoir for water than simply remove carbon dioxide from the atmosphere, an activity already carried out by plants without any effort from man.

The use of natural systems to remove the threat from excess nitrogen in aquifers and how to use natural systems for building up phosphate deposits has been more than adequately demonstrated.

Systems that have been achieving this for many years can be outlined.

The capture of storm overflow, with or without sewage and its containment prior to treatment and release, can fulfil the dual purpose of the storage of water within catchments for periods of drought.

Examples of all of these systems can be given and hopefully point the way to the use of carbon and catchments that can return water to the land and begin the stabilization of weather, particularly excess heat that is becoming a major problem and which will, if not attended to, bring about widespread famine and population displacement.

For those who wish to see a reduction in the use of energy *per se,* the amount of energy saved by using natural systems, presented as the monitored reduction obtained in working practice, will also be outlined.

The emission of gases by fossil fuel power generation is substantial and whether or not these contribute to global warming the reduction in such emissions has to be a reasonable objective from an air quality point of view. Wherever possible, therefore, the reduction in such emissions is also presented.

Before we move to that section however it is worth ending this part with the recent headlines from an EU report that provides a window on the future crisis from weather instability.

The report records the worst drought in five hundred years that has hit virtually the whole of Europe. Belgium, France, Germany, Hungary, Italy, Luxembourg, Moldova, the Netherlands, northern Serbia, Portugal, Romania, Spain, Ukraine and the UK are all in the grip of water shortage.

The change in climate has made for higher temperatures and increased the intensity of droughts. Night time temperatures that usually bring some respite from the heat are also increasing as the planet warms.

This is a dangerous development since the experience of Urban Heat Islands shows that inorganic materials, short of soil carbon and hence unable to store water, fall victim to severe temperature rises and act as storage radiators.

Yields of maize, soybean and sunflower are down by

16%, 15% and 12% compared to the previous five years because of water and heat stress. Wildfires have increased dramatically as ground and bushes dry and in Valencia (Spain) a fire of 20,000 hectares has been experienced.

The impact on population as well as environment is equally bad with thousands of people displaced and reportedly, hundreds dying of heat related conditions.

It is noted that there is a clear deficit of soil moisture yet the connection between this in terms of sensible and latent heat as experienced in towns and countryside and reflected in the temperature gradient between the two is not made or discussed.

Instead the comment: "More rapid adjustment by governments needed to reach net zero by 2050, professor says". The terms 'Nero' and 'Fiddling' come to mind.

If the degradation of soil continues and the term 'net zero' becomes a description of soil status then it will be the greatest disaster ever experienced by humanity.

The real questions from all of this, therefore, are as follows:

If the trees of Los Angeles can pump groundwater out of the soil into the sky and by doing so reduce the regional daytime temperature by 8°C, how do we put enough water and trees (plants) back into our catchments to bring about the same effect?

Can we at the same time reduce the emissions from power stations by using less power to treat sewage and effluent, which at its simplest produces an economic benefit in an era of expensive power?

Can we use soil and plants to manage our catchments in such a way that flooding is prevented?

Can we also store water in catchments and make it

available to farmers in periods of rainfall shortage?

Can we pump polluted aquifers and treat or capture the nitrogen and phosphorous contaminants in order to reuse the reclaimed water?

The examples given in the next section are all taken from existing operational systems that have answered the questions posed above in an affirmative manner.

Chapter 8

Sewage Treatment and Carbon Capture

Sewage by its nature is part of the human food cycle. As such it has a molecular formula expressed as: $C_{106}H_{180}O_{45}N_{16}P$

This is the formula used by engineers to calculate the energy requirements and other factors involved in the design of systems for treating sewage. Essentially it expresses the residue produced by the internal digestive system of the human body.

In simple terms it is saying that molecules and compounds that make up the material are composed of one hundred and six parts of carbon plus one hundred and eighty parts of hydrogen and so on up to phosphorous (one part).

The initial origin of these materials was in the plants we eat, which is of course taking these components out of the soil and rearranging them into something that is attractive for humans to consume.

The system at Othfresen was the first of its kind to be specifically employed for the treatment of sewage by natural methods. Previously sewage, or 'night soil' as it was euphemistically called, had been processed by combinations of sludge basins, rotating filters and other mechanical means that were energy intensive.

The makeup of wastewater is usually that of water loaded with dissolved organic substances together with polluting and sludgy substances which are decomposed by micro-organisms with the consumption of oxygen.

The wastewater flows are so big and highly polluted however that they would defeat the self-purifying processes of naturally flowing waters such as rivers, lakes and ponds.

By understanding the symbiotic action of soil, plants and bacteria however the self-purifying or decomposing processes can be made so compact and effective that they become readily feasible and operate with sustained and extremely high performance.

The monitoring that took place over a considerable period at Othfresen produced results whose significance only increases with time.

For the initial seven years the system served some 2,500 people and this then extended to 4,500 people for a further nine years. The objective initially had been to use the treated output from the system to maintain the water levels in the nature reserve.

As time went on, however, it became clear that the 'natural' method was treating the sewage to a quality that was superior in every way to the conventional system with which it was being compared.

Not only were the results for BOD and nitrogen (key parameters for sewage) exceeding the quality produced by the conventional system, but the bacterial and viral qualities of the treated water were exceptional.

Later work has shown that these systems are hugely

effective against all types of bacteria and virus and indeed the systems will manufacture new antibiotics in response to a change in sewage inputs.

A major characteristic, superior to conventional systems, was the ability to absorb a high degree of dilution by stormwater without a loss of functionality a property much in demand, but sadly lacking in today's sewage systems.

The monitoring carried out was on the simple basis of a fifty percent split of incoming sewage from the conventional stream to the reed beds, the remainder being sent to the conventional sewage works. In terms of energy comparison, therefore, the energy data was for the pumping of that volume of sewage to the reed bed facility.

The actual (man made) energy use of the reed bed was zero. Flow through the system was by simple gravity working on a small gradient through the beds. Soil/plant bacteria driven by sunlight provided the necessary action.

For the sixteen years of monitored operation the energy use and consequent emissions were as follows (coal fired power station, coal usage):

Reed Beds	Conventional works
Coal 17,000t	20,000,000t

Emissions Equivalent

CO_2 47,000Kg	53,000,000Kg
NO_2 70Kg	81,000Kg
SO_2 125Kg	142,000Kg

If one wishes to speak of carbon sequestration

then for this system a total of 5,500 tons of equivalent of CO_2 emissions was sequestered by mineralization processes in the soil.

Additionally, nitrogen was converted and sequestered into humic material and phosphorous was similarly mineralized into the soil of the reed beds.

With the present ideology of Net Zero then these figures meet all the criteria required for technology to overcome the present crisis. The huge reduction in energy requirements for treatment that as a result provide a major cut in emissions of greenhouse gases, the sequestration of carbon, nitrogen and phosphorous, which prevents the pollution of aquifers and rivers, are all beneficial effects irrespective of whether or not CO_2 is producing warming in the climate of the world.

The report produced after the visit of the Water Authorities contains interesting comments and recommendations. To paraphrase a number of sections:

"The Root Zone Method has the following benefits:
 – Environmentally acceptable offering considerable
   wildlife conservation potential.
 – Simple construction with no mechanical or
   electrical equipment.
 – Robust process able to withstand a wide range of
   operating conditions.
 – Low maintenance requirements.
 – Consistent effluent quality."

The author of the report came back from the visit "convinced that the RZ system was something which Britain's water authorities can't afford to ignore".

The capital cost of the Othfresen beds totalled £140,000, while annual operating costs come to less than

£6,000. These figures compare with the £1,000,000 capital and £60,000 operating costs of the nearby conventional works with which the comparison was being made (costs reported in pounds sterling).

"Add to these direct savings the other advantages of the RZ system over conventional sewage treatment – it dispenses with the need to find disposal outlets for sewage sludge, creates virtually no odours and has an environmentally friendly appearance – the explanation for the Water Authorities enthusiasm rapidly becomes clear."

This author is still waiting for a truthful explanation of why such an innovation was buried in favour of the construction of hundreds of fraudulent 'zero carbon – gravel systems' that have achieved virtually nothing over the years and now are mostly blocked and unusable.

Othfresen Inlet Areas as seen by staff of WRC in December 1984

Inlet channel and reed bed on the Southern works after 10 years of operation. Carbon sequestration was already in tons per annum. Phosphorous was entrapped and ammoniacal ammonia removed.

Interestingly, the costs of refurbishment are substantially higher than would be the replacement of a soil based system.

Gaseous Pathways

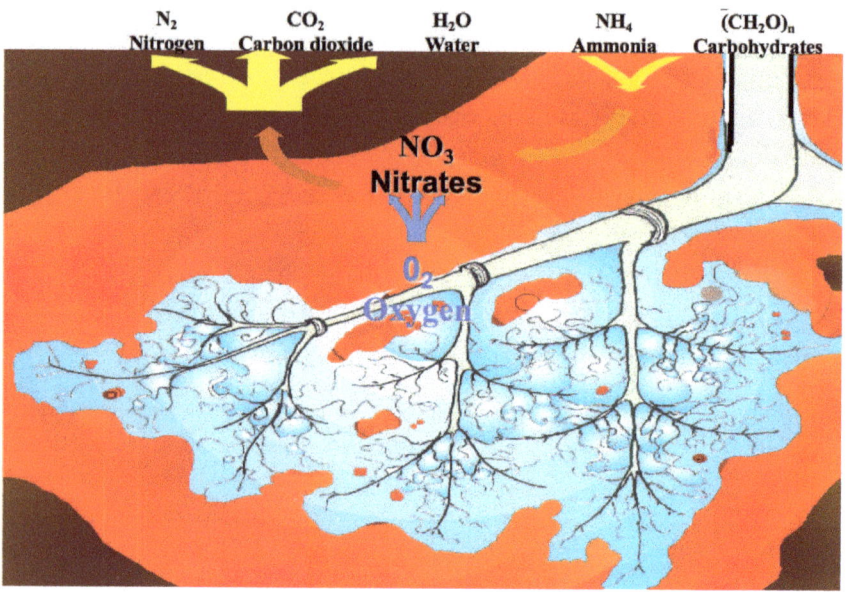

The roots of the plant (in this case a wetland phragmites) both add structure to the soil and enable soil (bacterial) respiration.

Chapter 9

Water

From the point of view of the thesis of this book, however, there is another factor implicit in the operation of soil based reed beds that adds enormously to all of the benefits mentioned.

In building such a system there is a commonality to the soil structure that arises from the interaction of a wetland plant and soil. Most people will be familiar with a household plant pot whereby the initial cutting occupies only a small part of the soil space.

With maturity, however, lifting the plant from the pot reveals a mass of rooted structure where soil and root are so intertwined that they become indistinguishable.

In the first instance the soil has its own primary structure and there is little contribution to this from the root ball of the plant. The porosity of the soil for watering is dependent on the micropores of the soil which contain small voids full of diffused air.

The clay component carries electrical charge, which

gives some structure and also binds water. The fibrous organic matter acts rather like girders in a building and maintains relatively open pore structures.

Excess watering of the soil displaces the air, neutralizes the clay charges and, over a period of time, destroys the organic fibre. The primary structure of the soil collapses, becomes impermeable and air can no longer penetrate in order to maintain aerobic conditions for soil bacteria. If the plant is a terrestrial species then it will rapidly fail if the soil remains flooded.

Contrasting this with a wetland species shows a major difference in that the roots, carrying their own oxygen with them, will continue to grow and spread and this time a secondary, stable, soil structure will be created. This secondary structure is a function of the roots rather than the soil and drastically alters the below ground situation of the plant.

Whereas a terrestrial soil will vary greatly in structure and water content depending on the percentage of clays, sand and organic matter that compose the soil, together with the variation in underlying strata, a secondary stabilized soil is more akin to a large sponge saturated with water.

For certain wetland species the pore space of the rooted soil may be as high as forty to fifty percent. Thus a terrestrial soil may be holding only forty tons of water per hectare in its top layer and available to plants, whereas a root zone system running to a depth of seventy centimetres over the hectare could hold up to three thousand tons of water per hectare and yet still appear to be a stable solid.

This incredible volume of available water can

therefore have a major impact on the water balance of a catchment.

To put this into context we should consider the following warning presented at a conference in the mid-nineties:

Seventeen countries in the Middle-East including Israel, Syria, Jordan, and Egypt, as well as South Africa, Pakistan, Southern India and Northern China, will face absolute water scarcity by twenty-twenty five.

Twenty four countries in sub-Saharan Africa will face severe economic water scarcity by twenty-twenty five.

Globally seventy percent of water taken from rivers or aquifers is used for irrigation (agriculture) and sixty to eighty-five percent of this is wasted.

In other words in the last century almost fifty percent of our land based water was finding its way into the sea and the situation since then has deteriorated.

Add to this the amount of water abstracted for drinking, which then transmits as sewage discharged down the rivers and into the sea, and one can begin to understand why the land is drying out.

The criticality of this is outlined in the recent report from the EU that shows quite clearly that the amount of water held within the soil of the land throughout the continent is falling substantially. It is reported also that the carbon content of soil is falling and yields of food crops are reduced. This crisis is being played out in many parts of the world.

The deadly trap that is beginning to close can be seen in the western sea board of America. Overuse of aquifers by extraction for agriculture has depleted them to such an extent that the low levels of water are out of reach of the tree roots and the cooling effect of the tree transpiration is no longer available.

The leaf stomata will close as the trees attempt to conserve their water. With the absence of the cooling effect beneath the canopy, the soil in which the trees are standing will itself dry out.

Without the removal of heat from the ground by the trees, the soil temperatures will continue to increase until the auto-ignition point of tree oils and resins is reached and the resultant conflagration will spread like the proverbial wildfire, which it is.

The rain experienced in day to day life arises both from the land via vegetation, soil and trees, and from the wind passing over the seas. When the sea-born clouds reach land, they lift up over the land surface and experience a cooling effect as they rise resulting in the water vapour condensing as rain.

As a result the coastal regions are blessed with rain. Now, with the tree and vegetation cover lost, the bare ground heats up and the cloud formations will rise on the convection currents produced and transport away from the coast.

The area will become dry and arid.

With the lack of water from the aquifer the ability to fight fires will be reduced and the water required for the considerable period of years needed for substantial tree growth will be absent. Reclamation becomes almost impossible.

Finally, this cumulative loss of water from the land being experienced in many parts of the world is in fact responsible for the majority of sea level rise forecasted erroneously as the result of glacial melt.

We can now, therefore, extract a common theme from the mass of information given so far. The large scale use of fertilizer, directing agriculture away from traditional methodologies, has resulted in the loss of organic carbon from soils.

The lack of carbon has severely reduced the water carrying capacity of the soil and in addition transported vast amounts of soil fines into rivers and drainage systems.

The loss of soil water has prevented the sensible heat from sunlight being converted into latent heat, and thus reducing the soil (ground) temperature.

Fields that employ traditional methods are seeing a ten degree temperature drop compared to fields operated agriculturally in the modern way and this reflects in higher air temperatures and drought in the agricultural areas.

The Urban Heat Island effect derived from brick or concrete which maintain an up to eight degree or higher temperature difference between city and country is now in danger of increasing and extending its range to dry soil in the arid countryside.

The dry, 'panned' soil discharges rainfall rapidly and overtops the drainage systems and rivers.

In England, the degree of overcrowding and continuous building on greenbelt land exacerbates the situation and regularly produces flood scenarios from convective rain, coupled with storm overflow from sewage works.

The problems are then compounded in that not only are houses saturated with muddy water but the water is carrying a high bacterial and viral load from the sewage works.

Chapter 10

Surface Water Problems

These problems of surface water arise from a number of situations touched on previously. In summary, rain, occurring frequently, or as a storm event, overloads the natural or man-made drainage systems and 'overtops' onto the land.

The Victorian idea of building well constructed and substantial sewers for industry and housing to carry away polluted water for treatment and discharge to rivers and estuaries was well thought out.

One component, however, now presents itself as a major problem. Rainwater, falling onto streets and houses, is also directed down into the underground system.

This combined 'foul' and surface water system places a major burden on the receiving treatment works whenever a major rainfall event occurs.

It is not widely recognized that water authorities are allowed to charge for rainwater falling on and discharging from industrial and commercial properties. Public sector buildings such as schools and hospitals are also charged substantial amounts for rainwater.

In financial terms there are charges for sewage and also a charge for treating rainwater.

Premises such as supermarkets having substantial carparks pay charges additional to sewage for the rainfall running off from their premises. The same applies to all non-domestic properties and adds up to a substantial cost.

If the volume of water reaching a treatment works becomes excessive, the Water Authorities have conditions placed on them allowing discharge of untreated materials into the nearest river or water course. These are the 'storm overflows', and number in excess of twenty thousand, that are responsible for the increasing amount of flooding in England.

In theory the substantial charges made upon customers for treating rainfall landing on their premises should be expended by the water authority on making provision to treat this water.

By the time the combined waters reach the works, however, there is no way of separating this volume from the sewage.

The large increase in population without the additional infrastructure and the greater intensity of storm events has produced a situation whereby regular flooding is occurring in many parts of the country.

This flooding is not just stormwater but water contaminated with sewage. Many people feel that discharges into housing areas and onto beaches are growing in number and severity.

The prosecutions that should occur arising from these events have been infrequent and inadequate and it seems that Water Authorities act without fear of the

Environment Agency.

When large organisations act with impunity and Agencies who should police their behaviour fail to do so, then it is time for the people who have been impacted to take action. Much is made of the 'Polluter Pays' principle, but experience shows that this is just empty rhetoric.

In future, crowd funding for Civil prosecutions should be used and that portion of Water Authority charges allocated to surface water (rainfall) discharges should be seized and used by the people to construct their own flood protection system.

The Weather reports will generally announce forthcoming storm events as 50 year or 100 year events and infrastructure design takes these calculated figures into account whenever a development is being undertaken.

Unfortunately, the modern phenomena of convective rain, bringing a rapid downpour of heavy rain and this coupled with higher volumes of sewage, are now producing flooding on a yearly basis and the volume of rain, together with the increased frequency, is now usually an underestimate of what can be expected.

There is no elasticity left in the system.

The situation requires a solution that both contains and then treats the contaminated water followed by slow release into adjoining ecosystems.

Such a requirement is familiar to those who design and operate the motorway networks.

A typical storm event on a motorway will produce fast moving flows of rainwater contaminated with Suspended Solids, Fuel, Poly Aromatic Hydrocarbons and

De-Icing Agents. Collectively these materials will prove toxic to aquatic life existing alongside motorways and some form of mitigation is therefore required.

A study some years ago by Imperial College compared the efficiency of various systems for containing and treating such run-offs.

One can see in the data reported that all of the conventional solutions provided suffer to a different degree from a failure to contain and/or treat the contaminated material exiting a motorway.

It is a staggering thought that a total of two thousand four hundred kilograms of solids per kilometre per year are generated by a motorway and these solids contain the toxic PAH materials that are devastating to aquatic life.

Motorway Data: Estimated load for Magor to Ilanwern section. Single "worst case" event generating 11,000 meter cubed of storm water.

Parameter	Load Kg/Km/yr	15Km length
Total Solids	2400Kg	36,000Kg
Total Suspended Solids	1500	22,500
Total Volatile Solids	740	11,100
Oil	126	1890
PAH	8103*	270gms

* Not verified – insufficient data

Above, solids, Oil and PAH in run-off from motorway based on average of 500mm p.a. rainfall. The figure in the right hand column is calculated from data in the second column.

The same team investigated the removal efficiences of types of interceptor and drainage system.

Below: Typical Containment Removal Efficiences for interceptor and drainage systems

Parameter	Sedimentation Tank	French Drain	Lagoon
TSS	52%	85%	92%
COD	35%	59%	54%
Oil	30%	70%	>70%*
PAH	<30%	70%	>70%*

* Not verified – insufficient data

With one or other of these systems the results tables demonstrate quite clearly   that with the best efforts available the amount of toxic materials escaping capture is high and persistent.

The lagoon and settlement tank may capture and drop out the solids, but the systems are impacted by the next storm event whereby the rapid influx of water simply ejects the entrained solids back into the environment.

Wetlands also suffer with the downward percolation of water soluble contaminants into the aquifer, and none of the systems offer substantial treatment of any of the contaminants listed.

The extra column of figures presented do however indicate the basis for an interesting case study in what could be achieved with the use of technology that is truly green.

To put things into perspective, the Newport area of S.Wales is home to the steel plant of llanwern. In the mid eighties the costs of effluent treatment were high. A coke oven was in production and the effluent was energetically costly to treat.

A research program was initiated to examine whether or not a biological treatment system could effectively treat the liquors. A pilot project was run under the auspices of RZG, who constructed a pilot plant of soil-based reed beds, and a two year study provided indications that a high standard of treatment could be obtained, thus reducing the energy cost.

The research was undertaken at the same time as the program for ICI at Billingham in the north of England and the successful outcomes of both projects resulted in the construction of the two most complex systems in the UK.

The llanwern system was constructed of local clay for the system base and sides and the top soil covering the clay strata was scraped off and placed within the clay basins. Local phragmites reeds were planted and the system was equipped with inlet and outlet pipework.

The methodology of enabling the symbiotic plant/soil/bacteria to adapt to the strong effluent and degrade the contaminants had, as with ICI, been worked out in the research phase and, once the system adapted, it continued successfully treating the effluent for over twenty years.

The first part of the system consisted of five hectares of reed beds with pipework that supplied liquor to just below the bed surface, which then flowed down through the soil and plant roots that were rich in bacteria.

The effluent contained oils, high ammonia levels and the toxic PAH material, and all had to be treated to certified levels to meet consent for discharge.

The clean liquor exiting the beds ran through a dedicated canal isolated from the surrounding vulnerable wetlands before discharging into the Bristol channel.

The system ran effectively for over twenty years before the closure of the coke ovens, when it was switched to treatment of much milder effluents.

Years later, the Welsh government proposed construction of a new section of the M4 motorway. This would pass from Magor along the underside of the steel works and over the river at Newport.

The gradient from Magor to the steel works was downhill for fifteen kilometres and calculations showed that a storm event would potentially generate a total of eleven thousand tons (m.cubed) of stormwater containing the contaminants outlined.

Unfortunately, the area into which this water would arrive was the Caldicott levels, an area which stands amongst the most sensitive environmental reserves in the UK.

A second problem was that the path of the motorway was directly through the reed bed effluent treatment plant for the steelworks, a plant that was still active.

Since this would therefore require the deconstruction and reconstruction of the reed beds in a location a short distance from the path of the road, but in a way that maintained continuity for the steelworks, it was decided to examine the possibility of reconstruction of a part of the beds into a containment and treatment system for both stormwater and the works effluent.

The total area of the first section was in the order of fifty thousand square meters split into five sections of approximately ten thousand meters each having a freeboard of just over a meter.

In effect, a storm event of eleven thousand meters cubed could be just accommodated into one section of the system. The system had been built to treat the PAH and other materials and there was no continuity between it and the surrounding Caldicott levels.

The study revealed that not only could the storm event be contained, but that it could all be treated to an exceptional standard prior to discharge into the tidal Bristol channel.

The motorway was, however, cancelled, but had it gone ahead then the implementation of the scheme described would have provided a major reduction in the environmental impact and in a way that could not be achieved with wetlands, lagoons and conventional civil engineering.

The same technique has been applied to sewage treatment and has demonstrated not only effective treatment for the normal sewage load for which it was designed, but also a capability to absorb and treat the shock loads that arrive with storm events.

This ability to capture and treat these excursions together with the capability to hold on to excess water until it is safe to discharge is now of primary importance in the catchments of this country.

The system as applied to sewage is illustrated opposite, next with readings obtained independently by the statutory authority.

With this system the incoming sewage arrives on

Performance

	Inlet	Dept of Env Sampling Results	% Reduction	Regulatory Standard (at DoE Sampling Location)
Secondary Treatment Parameters				
BOD (mg/t)	106	3.0	97.2.	20
TSS (mg/t)	1622	2.0	99.9	30
Tertiary Treatment Parameters				
Nitrogen (ammonia (mg/t)	17.3	0.5	97.1	2.0
Total Phosphorus (mg/t)	2.2	0.01	99.6	1.0
Total Coliform (MPN/100ml)	1.450,000	770	99.9	5000
Fecal Coliform (MPN/100ml)	1,160,000	260	99.9	1000

Design Parameters

Population:	1470
Population Equivalent:	1800
Average design flow:	3037 m³/day
Peak Flow capacity:	4555 m³/day
Extreme Peak Flow (storm water):	12,000 m³/day
BOD loading:	150 kg/day
BOD concentration:	26–160 mg/l

the left of the picture *(page 96)* into a series of sludge settling chambers. Under normal operation the liquor, minus the solids, moves forward to the vertical bed. Treatment begins as the liquor passes vertically down through the bed and this arrangement also allows any remaining solids to be trapped and mineralized on the bed surface.

From the vertical section the liquor moves forward under gravity and into and along the horizontal beds to finally exit and discharge into a pristine salmon river.

There is also a Storm Water Bed with a high freeboard that comes into play during storm situations. At these times the volume of combined rain and sewage water increases but the concentration of sewage requiring treatment falls. Unlike conventional sewage works the soil-based reed beds cope automatically with the changes in inputs.

It should also be noted that the results for all parameters exceed the statutory requirements at all times.

In parts of Europe there is contention about the pollution of aquifers with ammonia (nitrogen) and phosphorous, and this pollution which results from the use of fertilizers is now being used to justify the removal of farmers from their land.

It is obvious from the results of the system pictured that these two contaminants are effectively removed from the effluent passing through the system, and indeed far higher levels of ammonia have been successfully dealt with by this process.

The results for phosphorous are indicative of the fact

that this material can be sequestered into the soil of the reed beds. Experiments at the original Othfresen system and elsewhere show that high levels of phosphorous can be sequestered into specially prepared soils in order that the expected shortages of this material brought about by its over-extraction and misuse in fertilizer compounds could be avoided.

However, rather than mitigate the contaminant of aquifers by these materials, it appears that governments are calling for farming to cease and food production to be moved to S. America, onto land obtained by clearing rainforest.

(NOTE: As mentioned previously, the algal pollution affecting many parts of the Caribbean area is produced by the excessive amount of fertilizer already being used for such a purpose.)

As an alternative, systems such as these would be more than adequate for providing an abatement technology for individual farms both to provide clean water for irrigation by pumping from the aquifer and to prevent further contaminated water from penetrating into the aquifer.

It should also be possible to use aquifer water containing both ammonia and phosphorous as part of a methodology for returning to a regenerative agriculture, which puts carbon back into the soil and does away with fertilizers altogether.

To summarize, the table of design parameters shows a remarkable picture of a system that not only treats sewage to exceptional standards, but that can cope with storm events that are up to four times the design capacity.

This is way beyond anything achieved with conventional systems in this country, which are regularly discharging storm water heavily contaminated with sewage into built up areas during the winter season.

Typical Layout of Integrated System

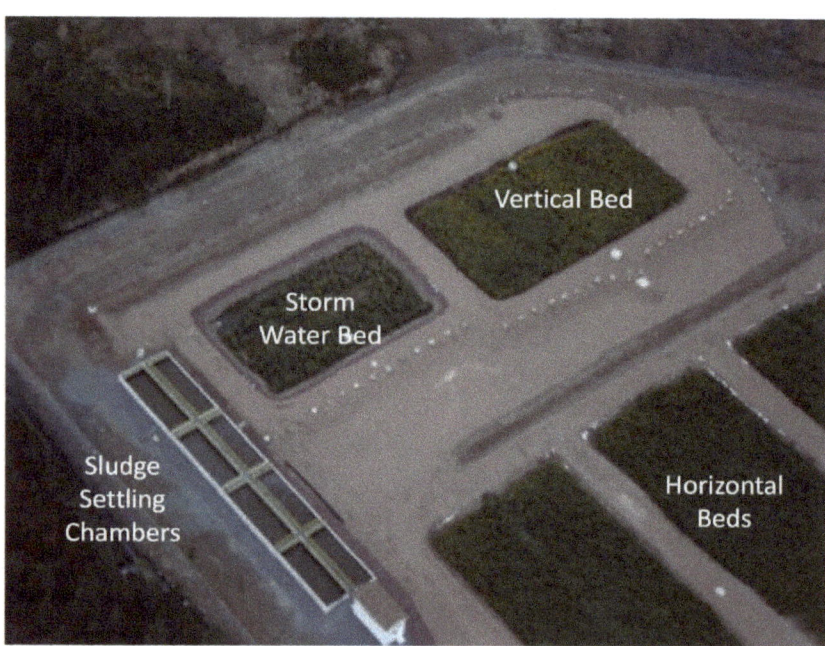

Canadian system discharging into salmon river.

Chapter 11

Energy and Emissions

The data obtained over many years from the systems at Othfresen, ICI and elsewhere demonstrate quite clearly that 'green technology' is capable of providing a great benefit to society.

Whilst disagreeing with the present paradigm of global warming, it is still possible to agree that reduction in emissions and the saving in energy are still objectives worth striving for.

The operation of Othfresen for just sixteen years resulted in a saving of fifty three million kilograms of carbon dioxide emissions, together with pro rata reduction in oxides of nitrogen and sulphur.

Additionally, with the capture of one million five hundred thousand kilograms of carbon into the soil, the whole experiment demonstrates a major route to addressing the perceived problems of today's world.

Following this up with the twenty five year experience of ICI, whereby almost two million tons of carbon dioxide emissions reduction was achieved over a twenty-five year period, and more than fifty thousand tons of carbon were sequestered into the soil, leads to the conclusion that with further research, natural techniques employing soil and its microbiology would provide a major benefit to mankind.

At the same time, the containment and treatment potential examined in the case of sewage systems and the S. Wales motorway storm events demonstrate also that the placement of such systems into catchments would not only mitigate harm from flooding, but also provide water for farmers in times of drought.

The ICI system alone could retain full functionality of treatment whilst holding fifteen thousand tons of water in the pore space of the soil.

Summary

The themes of this book are of course firstly that the use of fossil fuels is in no way responsible for the instability in weather presently being experienced and, secondly, that the real culprit is the reduction of carbon in agricultural soils which renders them incapable of holding water.

As a side issue, the impact fertilizers and worming agents are having is the extreme damage to the world's ecosystems and, in providing the incentive to industrial agriculture, are leaving a legacy of damaged soils that will take time and effort to repair.

This mistake has led to the UK being desperately badly served by 'Net Zero' policies that prevent the extraction and use of coal, oil and gas. Jobs have been deliberately sacrificed, export income lost and people are being exposed to the cold in winters whilst the population are held to ransom by less than friendly countries.

The true thermal effects impacting the planet are therefore a hydrogeological phenomenom, and it is the water cycles of the planet that need remediating.

Europe, together with the UK, have been blessed with climates that provide ample rainfall and, therefore, the full effects of weather instability have not yet become apparent.

The implementation of techniques based upon natural systems such as soil based reed beds and forestry would provide a number of high priority actions that are all beneficial to people and their communities.

Conclusions

The climate instability presently affecting the earth is not, as is reported by the media, a problem arising from the use of fossil fuels, but a two stage impact produced as a first step by the loss of carbon from soils as a result of bad agricultural practice.

The loss of carbon reduces soil to its inorganic components and of themselves these are unable to hold on to free water within the soil structure.

The heat transfer mechanism, i.e. the weather, of the planet is dependent upon the thermal properties of water as a heat transfer fluid.

A dry soil absorbs the radiant heat of the sun and its temperature rises. A wet soil transfers the heat to the water, which evaporates thereby cooling the soil.

Urban Heat Islands are produced by this combination

and contrast of hot inorganic building materials and the cooler areas of farm and forest. Ground and air temperatures can vary by as much as ten degrees within this scenario.

Dry soil is easily damaged by rainfall and washes off into channels and rivers. The remaining soil becomes 'panned' and water cannot then penetrate into the aquifer below.

Housing and other urban development add to the problem of aquifer isolation; crops and vegetables, which are unable to draw water from the soil, are reduced in yield and cannot in some cases be harvested from the soil.

Deforestation adds enormously to the problem since trees cool the ground by transfering large amounts of water into the upper atmosphere, where the water vapour cools and forms into clouds.

The hypothesis put forward here is up for discussion, of course. However, in refusing to discuss the present climate change dogma, the proponents of it are laying the ground for further colossal damage to the world's ecosystem and the humanity that depends on it.

The current dogma is nihilistic. A deliberate choice by government not to earn money by exporting oil, coal or gas. Coalfields throughout the country are abandoned and idle and the communities who live there have been blighted. Potential jobs in Cumbria, created by mining and exporting coal, turned down for an unproven consensus arriving from a policy that dictates science.

It is well past the time for those who really rule our country to be called to account. The global warming narrative is a fake.

In the eighteenth century the French Academy of Sciences declared that "rocks do not fall from space, there is no such thing as a meteorite". Ten years later the Academy reversed its decision.

As for climate change, the BBC has declared that "the science is settled". I hope that, one hundred years from its inception, this statement becomes its epitaph.

ICI Billingham

Input = 3,000m³ /day, up to 30,000mg/l COD phenols, amines, detergents etc. 30+yrs operation estimated 17,000 tonnes of carbon sequestrated. Worldwide interest. Water holding capacity over 5 hectares equal to 15,000 tons.

CATCHMENTS AND CARBON

About the Author

The author is a qualified scientist who has worked in the field of fuel and energy before moving to anti-desertification and soil remediation. Projects in this area have been carried out in all parts of the world and provided an extensive knowledge of soils and ecosystems.

Above all believing science has a duty to explore and develop new theories and methodologies for solving the problems manifest in the world – the case is never closed.

The statement put out by the BBC that, for global warming, "the science is settled", is therefore a sure indication that a strategy other than scientific guidance is being undertaken.

Additionally, the concept of education of children in the impact of global warming is a sure sign that indoctrination rather than education is the objective.

As a consequence of these two factors, this book puts forward a reply based upon a holistic argument that actually seeks to explain and validate the role of water and soil carbon in the unfolding disasters produced by climate instability.

The 'Net Zero' policy based upon a false paradigm will only exacerbate the scale of the crisis whilst additionally bringing economic collapse to the UK economy. Nobody in this country voted to stop oil production, keep our coal mines closed or destroy our power stations.

A referendum expressing the will of the people, an instrument detested by the political class, is required to enable the nation to decide on the way forward.

*

Contact:

Email: lois@loiscaborn-institute.com

Website: https://loiscaborn-institute.com

www.ingramcontent.com/pod-product-compliance
Lightning Source LLC
Chambersburg PA
CBHW041646200526
45172CB00022BA/1280